LEBEN UND WERK

Man tut nicht etwas, nur weil es einem Spaß macht,
sondern weil es richtig ist.
Ludwig Mies van der Rohe

DAS „ARCHITEKTONISCHE GEWISSEN DES JAHRHUNDERTS“

Ludwig Mies van der Rohe hat wie kein zweiter Architekt das Bild des 20. Jahrhunderts geprägt. Es gibt kaum eine bedeutende Stadt, in der nicht ein Bau nach seinem Vorbild stünde, und erst recht keinen Architekten, der nicht seine eigene Position gegen-

Mies vor dem Modell der Crown Hall um 1955

LIFE AND WORK

You don't do something only because you enjoy it,
you do it because it is right.
Ludwig Mies van der Rohe

THE "ARCHITECTURAL CONSCIENCE OF THE CENTURY"

Ludwig Mies van der Rohe has dominated the visual identity of the 20th century more than any other architect. There is hardly any major city that does not have a building modelled on his work, and every architect feels compelled to define his position

Mies in front of the model of the Crown Hall around 1955

über Mies' asketischer Bauauffassung definieren musste. Seine bis aufs Äußerste reduzierten Stahl- und Glasbauten kann man bewundern oder ablehnen - aber ihren Absolutheitsanspruch nicht ignorieren.

Dabei spiegeln die Bauten, die sich heute mit dem Namen Mies van der Rohe verbinden, hauptsächlich die zweite Hälfte seines Lebenswerks wider, nämlich die nach seiner Emigration in die USA 1938.

Von seinem 19. bis zu seinem 52. Lebensjahr lebte und arbeitete Mies in Berlin. Hier verbrachte er seine Lehrjahre als Architekt, hier baute er mehrere Landhäuser, und hier entstanden die bahnbrechenden Ideen und Entwürfe, auf denen seine späteren Werke basieren. Und nicht zuletzt schuf er für Berlin das Schluss- und Glanzstück seines Œuvres, die Neue Nationalgalerie. Fast alle seine Berliner und Potsdamer Werke sind erhalten und noch heute zu besichtigen.

JUGEND UND ANFÄNGERJAHRE

Geboren wurde Ludwig Maria Michael Mies am 27. März 1886 in Aachen, als jüngstes von vier Geschwistern. Erst um 1920 fügte er seinem Namen den Geburtsnamen seiner Mutter, Rohe, und das niederdeutsche „van der" an. Seine Freunde, Bekannten und Mitarbeiter nannten ihn stets nur „Mies".

Mies war ein „Baumeister" im besten Sinne, ein Mann der Praxis, der lieber anpackte, als zu theoretisieren. Architektur hat er nie studiert. Im Betrieb seines Vaters, der als Steinmetzmeister auf Kamineinfassungen und später auf Grabsteine spezialisiert war, entwickelte er schon früh ein feines Gespür für Material und Respekt vor handwerklichem Können. Bei einer Baufirma lernte er das Maurerhandwerk, später arbeitete er als Zeichner in einem Stuckgeschäft und in zwei Architekturbüros. Als ihm ein Kollege riet, nach Berlin zu gehen, bewarb er sich auf eine Zeitungsanzeige und erhielt 1905 eine Stelle als Zeichner im Städtischen Bauamt Rixdorf (Neukölln), wo er die Wandverkleidung für den Plenarsaal des Rathauses ausarbeitete. Noch im selben Jahr wechselte er ins neu eröffnete Architekturbüro von Bruno Paul. Der berühmte Möbeldesigner (und Simplicissimus-Karikaturist) war gerade zum Direktor der Berliner Kunstgewerbeschule berufen worden, und dort schrieb sich auch Mies für kurze Zeit als Student ein.

Über einen Kollegen im Büro Paul lernte Mies den Philosophieprofessor Riehl und seine Frau kennen, die mit einem Bauauftrag einen jungen Architekten fördern wollten - allerdings keinen blutigen Anfänger. Dennoch bekam Mies den Auftrag. Zeitlebens hat

in relation to Mies' ascetic approach to building. You can admire or reject his buildings of steel and glass, but you cannot ignore their claim to absoluteness.

But the buildings which are now linked with the name of Mies van der Rohe only reflect the second half of his life after his emigration to the USA in 1938. Mies lived and worked in Berlin from the age of 19 to the age of 52. It was here that he received his training as an architect and built several country houses, and it was here that he developed his revolutionary ideas and designs which formed the basis for his later works. And last but not least, it was for Berlin that he created the masterpiece that crowned his work, the New National Gallery. Almost all of his works in Berlin and Potsdam are still preserved and can still be viewed today.

YOUTH AND EARLY YEARS

Ludwig Maria Michael Mies was born in Aachen on 27th March 1886 as the youngest of four children. It was only around 1920 that he added his mother's maiden name, Rohe, and the Low German form "van der" to his name. His friends, acquaintances and staff only called him "Mies".

Mies was a "master builder" in the best sense, a man who liked to get on with the work rather than indulging in theory. He never studied architecture. In the business run by his father, who was a master stone mason who initially specialised in chimney-pieces and later in gravestones, he developed an early sensitivity to material and a respect for skillful craftsmanship. He learned bricklaying in a construction firm, and later he worked as a draughtsman for a stucco company and two architectural firms. When a colleague advised him to go to Berlin, he applied for a post advertised in a newspaper and in 1905 he started work as a draughtsman for the municipal building authority of Rixdorf (=Neukölln), where he worked out the details of the wall cladding for the plenary chamber of the town hall. In the same year he moved to a job with the newly opened architectural office of Bruno Paul. The famous furniture designer (and caricature artist for Simplicissimus) had just been appointed as the director of the School of Design in Berlin, and for a short time Mies enrolled as a student there.

Through a colleague in Bruno Paul's office, Mies got to know the Philosophy professor Riehl and his wife who wanted to give a commission to a young architect to help his career - but not to a complete beginner. Nevertheless, Mies got the commission. Throughout his life, Mies impressed his clients with his calm and confident, yet unassuming manner.

er seine Klienten durch seine ruhige und selbstbewusste, dabei aber ganz uneitle Art beeindruckt.
Mit ihrem Haus, das keinesfalls wie ein Erstlingswerk wirkt, waren die Riehls so zufrieden, dass sie Mies eine mehrwöchige Studienreise nach Italien finanzierten und ihn in ihren Freundeskreis aufnahmen, in dem Leute wie Walther Rathenau oder der Kunsthistoriker Heinrich Wölfflin verkehrten. Hier erschloss sich ihm sein späterer Kundenkreis.
1908 wechselte Mies ins frisch eröffnete Büro des damals berühmtesten Architekten Deutschlands, Peter Behrens, der wie Bruno Paul vom Kunsthandwerk kam und als künstlerischer Berater des AEG-Konzerns für dessen gesamtes Erscheinungsbild verantwortlich war – vom Firmenlogo über das Produktdesign bis hin zur Architektur. Zunächst arbeitete Mies unter Behrens' jungem Büroleiter Walter Gropius (dem späteren Bauhaus-Gründer) an der Kleinmotorenfabrik in der Voltastraße (Wedding) und an der berühmt gewordenen Turbinenfabrik in der Huttenstraße (Moabit).
Auf eigene Rechnung beteiligte er sich am Wettbewerb für ein Bismarck-Denkmal bei Bingen – er entwarf einen riesigen, symmetrisch angelegten Bau hoch über dem Rheintal – und baute sein zweites Haus, die streng klassizistische Villa Perls in Zehlendorf.
Zurück in Behrens' Büro, leitete Mies 1911/12 den Bau der Deutschen Botschaft in St. Petersburg, eines neoklassizistischen Monumentalbaus, den er angeblich überhaupt nicht mochte, und betreute ein Landhausprojekt für das Kunstsammler-Ehepaar Kröller/Müller bei Den Haag. Weil Frau Kröller jedoch mit Behrens' Entwurf nicht einverstanden war, beauftragte sie Mies mit einem Gegenvorschlag. Den Sommer über arbeitete er in jenem Atelier, in dem die stattliche Van-Gogh-Sammlung der Kröllers hing. „So wurde ich ein Van-Gogh-Experte, ohne es zu wollen." Doch auch Mies' Entwurf einer weit ausgreifenden, streng klassizistischen Villa wurde schließlich abgelehnt.

VORBILDER UND EINFLÜSSE

Prägend wirkte zunächst Peter Behrens, der für die AEG wegweisend moderne Stahl- und Glas-Bauten entwarf, in seinen übrigen Bauten aber einen monumentalen Neoklassizismus pflegte. Behrens verehrte Karl Friedrich Schinkel, den man damals wegen seiner konstruktiven Klarheit als „modernen" Architekten wieder entdeckte, und besuchte mit seinen jungen Mitarbeitern oft die Berliner Schinkel-Bauten. Schinkels strenge Formsprache beeinflusste Mies stark.
Während seines Hollandaufenthalts lernte Mies das Werk Hendrik Petrus Berlages kennen und schätzen: Die Klarheit der Konstruk-

The Riehls were so pleased with their house, which certainly does not seem like the first house built by a beginner, that they paid for him to go to Italy for several weeks on a study trip and took him into their circle of friends, which included people like the industrialist and later Foreign Minister Walther Rathenau and the art historian Heinrich Wölfflin. This was where he got to know many later clients.
In 1908 Mies moved to a job in the office of the most famous architect in Germany at that time, Peter Behrens, who had his roots in Design like Bruno Paul and was the artistic consultant for the AEG group of companies with responsibility for its entire visual identity – from the company logo to the product design and the architecture. At first, Mies worked under Behrens' office manager Walter Gropius (the later founder of the Bauhaus) on the small motor factory in Voltastrasse (Wedding) and the famous turbine factory in Huttenstrasse (Moabit).
On his own account he took part in the competition for a Bismarck monument near Bingen – he designed an enormous symmetrical building high above the Rhine valley – and built his second house, the strictly Classical Villa Perls in Zehlendorf.
Back in Behrens' office, Mies directed the construction of the German embassy in St. Petersburg in 1911/12, a monumental neo-Classical building which he allegedly did not like at all, and he supervised a country house project for the married art collector couple Kröller/Müller near The Hague. But Mrs. Kröller was not satisfied with Behrens' design, so she commissioned Mies to present an alternative design. He worked through the summer in the studio which contained the Kröllers' splendid Van Gogh collection. "I thus became an expert on Van Gogh without wanting to." But Mies' design of a strictly Classical villa with a generous floor layout was eventually rejected.

MODELS AND INFLUENCE

Mies was initially influenced by Peter Behrens, who designed progressive modern steel and glass buildings for AEG but had a monumental neo-Classical style in his other buildings. Behrens admired Karl Friedrich Schinkel, who had recently been rediscovered as a "modern" architect because of the clarity of his structural design, and often visited Schinkel's buildings in Berlin with the younger members of his staff. Schinkel's strict language of forms had a strong influence on Mies.
During his time in Holland, Mies came to know and appreciate the work of Hendrik Petrus Berlage: the clarity of the structural design – "honest to the core" (Mies) and unburdened by history –

AEG
FABRIK

tion – „ehrlich bis auf die Knochen" (Mies) und ohne geschichtlichen Ballast – beeindruckte ihn ungemein. Berlage, als dessen Hauptwerk die Amsterdamer Börse gilt, strebte nach einem „objektiven", universalen Stil, als dessen Grundlage er eine auf geometrischen Proportionen gründende „Ordnung" ansah.
Einen wichtigen Einfluss übte schließlich auch der Amerikaner Frank Lloyd Wright aus, dessen Wohnhausarchitektur durch eine Ausstellung 1910 in Berlin bekannt wurde. Die fließenden Räume und freien Grundrisse in Mies' Arbeiten der 20er Jahre wären ohne Wrights Anregungen nicht denkbar.

ALS FREIER ARCHITEKT

1913 eröffnete Mies ein eigenes Architekturbüro in Steglitz und heiratete Ada Bruhn, eine Industriellentochter aus wohlhabendem Hause, die er bei den Riehls kennen gelernt hatte. Innerhalb von drei Jahren wuchs die Familie um drei Töchter. 1915 zogen sie in den zweiten Stock eines spätklassizistischen Mietshauses in Tiergarten (Am Karlsbad 24), das heute nicht mehr steht. Mies war allerdings weder Familienmensch noch treuer Ehemann, und es gibt Zeugnisse für große Ehekräche und dramatische Szenen. Um 1920 zog Ada mit den Töchtern aus, doch die Ehe wurde bis zu Adas Tod 1951 nicht geschieden. Bis zur Emigration 1938 lebte und arbeitete Mies Am Karlsbad.
Nur drei Landhäuser konnte Mies bauen, ehe er 1915 zum Militär eingezogen wurde. Als Schreiber diente er lange Zeit in einem Bataillonsbüro in Berlin, das Kriegsende erlebte er in einer Straßenbaukompanie in Rumänien.

IN DER SPEERSPITZE DER MODERNE

An den politischen Umwälzungen nach 1918 nahm Mies anscheinend keinen Anteil – wie er sich überhaupt zeitlebens nicht um Politik kümmerte. Künstlerisch jedoch wandelte er sich in den ersten Nachkriegsjahren zum Revolutionär. Mit fünf utopischen Entwürfen setzte er sich binnen weniger Jahre an die Spitze der architektonischen Avantgarde.
Allerdings schätzte Mies das gute Leben – er kleidete sich stets sorgfältig, rauchte teure Zigarren und seine Trinkfestigkeit war berüchtigt. Dies ließ sich nur durch Bauaufträge finanzieren, und so belieferte er noch bis Mitte der 20er Jahre seine zahlungskräftige Kundschaft mit neoklassizistischen Landhäusern. Es sind „ehrliche" und bestechend sorgfältig gearbeitete Bauten, wenn sie auch in erheblichem Gegensatz zu Mies' radikalen Mani-

◁ Die AEG-Turbinenfabrik von Peter Behrens in Moabit

greatly impressed him. Berlage, whose main opus is the stock market in Amsterdam, strove for an "objective", universal style, and he regarded an "order" based on geometric proportions as the basis of such a style.
Another important influence was the American Frank Lloyd Wright, whose residential architecture became known in Berlin through an exhibition in 1910. The flowing spaces and open floor plans in Mies' work in the 1920s are unthinkable without Wright's influence.

A FREELANCE ARCHITECT

In 1913 Mies opened his own architectural bureau in Steglitz and married Ada Bruhn, the daughter of a wealthy industrialist, whom he got to know through the Riehls. Within three years the family grew by three daughters. In 1915 they moved into the second floor of a late Classical block of rented flats in Tiergarten (Am Karlsbad 24), which no longer exists today. But Mies was not the family type, nor was he a faithful husband, and there are reports of violent rows and dramatic scenes. Around 1920 Ada moved out with their daughters, but the marriage was not divorced up to Ada's death in 1951. Until his emigration in 1938, Mies lived and worked in Am Karlsbad.
Mies was only able to build three country houses before he was conscripted to military service in 1915. For a long time he worked as a writer in a battalion office in Berlin, and he saw the end of the war in a road-building company in Romania.

A TRAILBLAZER FOR MODERNISM

Mies apparently did not participate in the political upheaval after 1918 – in fact he did not involve himself in politics as long as he lived. In his artistic approach, however, he became a revolutionary in the early post-war years. With five utopian designs he became a figurehead of the architectural avant garde within a few years.
But Mies loved the good life – he always dressed carefully, smoked expensive cigars and was known for his ability to "hold his drink". He could only pay for this life style by accepting building commissions, so up to the mid-1920s he built neo-Classical country houses for his wealthy clientele. They were "honest" and strikingly well-built buildings, even though they were in stark contrast to Mies' radical manifestos in that period. It was only after inflation had been overcome and money was finally available

◁ The AEG turbine factory by Peter Behrens in Moabit

M.SPITZ
Ankauf
von
Juwelen
Hygien-
Brillanten
GOLD
JUWELEN

festen aus jener Zeit stehen. Erst als nach überstandener Inflation endlich Geld für avantgardistische Bauprojekte vorhanden war, endete Mies' „künstlerisches Doppelleben".

Den Anfang der fünf Ideal-Entwürfe, mit denen er die prinzipiellen Möglichkeiten der neuen Materialien und der Konstruktion auslotete, machte 1922 sein Wettbewerbsbeitrag zum „Turmhaus" an der Friedrichstraße: ein ganz mit Glas ummanteltes Hochhaus, das auf Zeichnungen und Collagen durch seine Monumentalität und Kompromisslosigkeit besticht. Zudem verzichtete Mies auf einen bis dahin essentiellen Bestandteil jeder Architektur: die Fassade.

Einige Monate später präsentierte er einen zweiten Hochhausentwurf: schlanker, um die Hälfte höher und auf gerundetem, amorphem Grundriss. Mies hat stets betont, dass die Grundrisse nicht beliebig gewählt, sondern das Ergebnis zahlreicher Versuche an Glasmodellen waren, um ein optimales Spiel der Reflexe zu erreichen.

Mit dem „Bürohaus in Eisenbeton" stellte Mies einen achtgeschossigen, flach gedeckten Stahlbetonbau vor. Die einzelnen Geschossplatten bilden Brüstungen, zwischen denen durchlaufende Fensterbänder sitzen – eine heute ganz selbstverständliche Konstruktionsweise.

Mit dem „Landhaus in Eisenbeton" und dem „Landhaus in Backstein" entwickelte Mies schließlich zwei flach gedeckte, weit in die Umgebung ausgreifende Wohnhäuser, deren frei stehende Wände keine Zimmer oder Raumbereiche mehr umschließen, sondern allein den „Hausorganismus" gliedern.

Die Entwürfe, denen er knappe, bekenntnishafte Texte beigab, veröffentlichte er in Zeitschriften und auf Ausstellungen, die er zum Teil, etwa für die „radikale" „Novembergruppe", selbst organisierte. Mit dem Allround-Künstler Hans Richter und den Wortführern der Konstruktivisten, El Lissitzky, und der De-Stijl-Bewegung, Theo van Doesburg, betrieb Mies seit 1923 die Zeitschrift „G – Material für elementare Gestaltung".

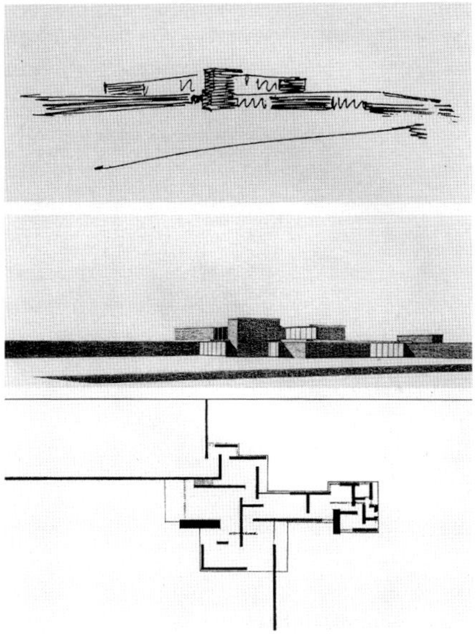

„Landhaus in Backstein" (Entwurf 1922)
"Country house of brick" (design 1922)

for avant garde construction projects that Mies ended his "artistic double life". The first of the five ideal designs in which he explored the fundamental possibilities of new materials and structural design was his competition entry for the "tower building" in Friedrichstrasse in 1922: a tall building completely enclosed in glass which, in drawings and collages, is impressive in its monumentality and lack of compromise. And Mies omitted an element which until then was an essential part of all architecture: the facade.

Some months later he presented his second design for a tower building: more slender, half as high again and on a rounded, amorphous ground plan. Mies always emphasised that the floor plans were not arbitrary, but rather the outcome of numerous experiments on glass models to achieve the best possible reflective properties.

With the "office building of reinforced concrete" Mies presented a flat-roofed eight-storey reinforced concrete building. The floor slabs of the individual storeys form parapets between which there are continuous window bands – a structural feature which is taken for granted today.

With the "country house of reinforced concrete" and the "country house of brick" Mies finally developed two flat-roofed residential buildings with an expansive ground plan and free-standing walls which no longer enclose rooms or spatial areas, but simply structure the "house organism".

The designs, which he accompanied with brief, confession-type texts, were published in periodicals and at exhibitions, some of which he organised himself, e.g. for the "radical" "November group". Together with the all-round artist Hans Richter and the leading proponents of Constructivism, El Lissitzky, and the De-Stijl movement, Theo van Doesburg, Mies published the periodical "G – Material für elementare Gestaltung" (G – material for elementary design) from 1923.

In fact, although Mies was always cautious and slow to put pen to paper, in these years he became one of the leading propagandists

◁ Wettbewerbsentwurf für das Turmhaus Friedrichstraße (Collage von 1922)

◁ *Competition design for Friedrichstrasse tower building (collage of 1922)*

Überhaupt wurde der stets bedächtige und schreibfaule Mies in diesen Jahren zum Propagandisten des Neuen Bauens. Er publizierte Artikel, die er allerdings meist zu spät und wesentlich kürzer als abgesprochen lieferte, hielt Vorträge und engagierte sich in Künstlervereinigungen. 1924 gehörte er zu den Mitbegründern der Architektengruppe „Der Ring", deren Mitglieder sich oft in seiner Bürowohnung trafen, 1926 wurde er Vizepräsident – und damit wichtigster Mann – des Deutschen Werkbundes, der damals führenden Künstlervereinigung.

ZU NEUEN UFERN

Nach dem Ersten Weltkrieg erfasste die europäischen Künstler eine enorme Aufbruchsstimmung, mussten doch nach dem Zusammenbruch der alten Ordnung neue Wege gefunden werden. Nach der Überwindung des Historismus galt es, die Beziehung zwischen Konstruktion und Form, zwischen Zweck und Wert neu

of new building. He published articles, although he usually submitted them after the deadline and shorter than agreed, he gave lectures and he became involved in artistic associations. In 1924 he was one of the co-founders of the architects' group "The Ring", whose members often met in his office-apartment, and in 1926 he became the vice-president – and thus the most important member – of the German Werkbund, the leading artists' association of the time.

A NEW START

After the First World War, European artists experienced an enormous sense of new beginnings because the old order had collapsed and new ways had to be discovered. After Historicism had been overcome, the relationship between structure and form and between purpose and value had to be redefined. The goal was no less than the "renewal of architecture".

Das „Bürohaus in Eisenbeton" aus der Zeitschrift „G" I/1923

The "reinforced concrete office building" from the magazine "G" I/1923

Jede ästhetische Spekulation,
jede Doktrin,
und jeden Formalismus
lehnen wir ab.

Baukunst ist raumgefaßter Zeitwille.
Lebendig. Wechselnd. Neu.

Nicht das Gestern, nicht das Morgen, nur das Heute ist formbar.
Nur dieses Bauen gestaltet.

Gestaltet die Form aus dem Wesen der Aufgabe mit den Mitteln unserer Zeit.

Das ist unsere Arbeit.

BÜROHAUS

Das Bürohaus ist ein Haus der Arbeit der Organisation der Klarheit der Ökonomie.
Helle weite Arbeitsräume, übersichtlich, ungeteilt, nur gegliedert wie der Organismus des Betriebes. Größter Effekt mit geringstem Aufwand an Mitteln.

Die Materialien sind Beton Eisen Glas.

Eisenbetonbauten sind ihrem Wesen nach Skelettbauten. Keine Teigwaren noch Panzertürme. Bei tragender Binderkonstruktion eine nichttragende Wand. Also Haut- und Knochenbauten.

Die zweckmäßigste Einteilung der Arbeitsplätze war für die Raumtiefe maßgebend; diese beträgt 16 m. Ein zweistieliger Rahmen von 8 m Spannweite mit beiderseitiger Konsolauskragung von 4 m Länge wurde als das ökonomischste Konstruktionsprinzip ermittelt. Die Binderentfernung beträgt 5 m. Dieses Bindersystem trägt die Deckenplatte, die am Ende der Kragarme senkrecht hochgewinkelt Außenhaut wird und als Rückwand der Regale dient, die aus dem Rauminnern der Übersichtlichkeit wegen in die Außenwände verlegt wurden. Über den 2 m hohen Regalen liegt ein bis zur Decke reichendes durchlaufendes Fensterband.

Berlin, Mai 1923 Mies v. d. Rohe

zu definieren. Das Ziel war kein geringeres als die „Erneuerung der Baukunst".
Die Lösung sah Mies allein in einer strikten Rationalität. Schroff lehnte er jede Ästhetik ab und konzentrierte sich ausschließlich auf die Konstruktion. „Die Form ist nicht das Ziel, sondern nur das Resultat unserer Arbeit. Es gibt keine Form an sich." „Baukunst" definierte er zu dieser Zeit als „raumgefassten Zeitwillen", und folgerichtig akzeptierte er uneingeschränkt die Technik und die neuen Materialien.
So konsequent wie seine radikalen Schriften waren seine Entwürfe allerdings nicht. Die Hochhausprojekte zeigen deutlich Mies' Faszination für die Ästhetik des Skelettbaus, und im „Bürohaus aus Stahlbeton" ließ er die Geschosse - konstruktiv unnötig - vorkragen. Um die Mitte der 20er Jahre rückte Mies dann auch verbal vom unbedingten Materialismus ab und definierte Bauen nun als „räumlichen Vollzug geistiger Entscheidungen".

MEILENSTEINE DER MODERNE

1925 endlich konnte Mies seinen Worten und Entwürfen Taten folgen lassen: In Guben und Krefeld baute er drei kubische, in ihrer Kargheit beeindruckende Backsteinvillen; in Berlin entwarf er das berühmt gewordene Denkmal für Karl Liebknecht und Rosa Luxemburg und seinen einzigen Beitrag zum sozialen Wohnungsbau: vier Wohnblöcke in der Afrikanischen Straße.
Endgültig in die Reihe der weltweit führenden Architekten stieg Mies mit drei Projekten Ende der 20er Jahre auf: Zunächst übernahm er für den Werkbund die Gesamtplanung der Ausstellung „Die Wohnung" und der Versuchssiedlung Weißenhof in Stuttgart, die der modernen Architektur schließlich zum entscheidenden Durchbruch verhalf.
Mit sicherem Gespür versammelte Mies die aus heutiger Sicht bedeutendsten Architekten des Neuen Bauens, und obwohl er den Beteiligten weite Freiheiten ließ, zeigt die Siedlung ein erstaunlich einheitliches Bild: Die neue Bewegung hatte zu einem verbindenden „internationalen Stil" gefunden. Bei der Gesamtanlage durchbrach Mies das Dogma des Zeilenbaus und gruppierte die 21 Bauten frei entlang einer gekurvten Straße am Hang. Sich selbst behielt er den größten Wohnblock vor, einen Stahlskelettbau, in dem die nicht-tragenden Wände gemäß den Bedürfnissen und Wünschen der Bewohner angeordnet werden konnten.
1929 entwarf Mies als künstlerischer Leiter des deutschen Beitrags zur Weltausstellung in Barcelona einen reinen Repräsentationsbau, von dem sein alter Lehrmeister Peter Behrens prophetisch sagte: „Diesen Pavillon wird man einmal als das schönste

For Mies, the solution could only lie in a strict rationality. He firmly rejected aesthetics and concentrated exclusively on structural design. "The form is not the goal, it is merely the result of our work. There is no such thing as form in itself." In this period he defined "architecture" as a "spatial expression of the will of the age", and it was therefore logical that he accepted technology and the new materials without any restrictions.
But his designs were not as consistent as his radical writings. The tower building projects clearly show his fascination with the aesthetics of skeleton frame construction, and in the "office building of reinforced concrete" he designed storeys which projected out from the building - which was structurally unnecessary. Around the middle of the 1920s Mies also departed from unconditional materialism in his statements, and he now defined building as the "spatial implementation of mental decisions".

MILESTONES OF THE MODERN AGE

In 1925, Mies was finally able to put his words and designs into practice: in Guben and Krefeld he built three cubic villas which were impressive in their simplicity, in Berlin he designed the monument for Karl Liebknecht and Rosa Luxemburg, which later became famous, and his only contribution to social housing, three residential buildings in Afrikanische Strasse.
Mies finally became one of the world's leading architects with three projects at the end of the 1920s. First, on behalf of the Werkbund, he took on the overall planning of the exhibition "Die Wohnung" (The apartment) and Weissenhof experimental estate in Stuttgart, which finally brought the decisive breakthrough for modern architecture.
With a sure instinct, Mies gathered the architects of New Building who are now regarded as the leading figures, and although he gave the participants a large degree of freedom, the estate had a surprisingly consistent appearance: the new movement had found a unifying "international style". In the overall complex, Mies departed from the dogma of elongated buildings and grouped the 21 buildings in a loose pattern along a curved hillside road. He reserved the largest residential block for himself and designed a steel frame structure in which the non-structural walls could be arranged according to the needs and wishes of the residents.
In 1929, as the artistic director of the German contribution to the International Exhibition in Barcelona, Mies designed a building with a purely representational character of which his former teacher, Peter Behrens prophetically said: "This pavilion will even-

Der „Barcelona-Pavillon"

The "Barcelona pavilion"

Bauwerk des 20. Jahrhunderts bezeichnen." Erstmals verwirklichte Mies hier sein Ideal des frei fließenden Raumes. Unter dem scheinbar schwebenden Flachdach, das auf acht regelmäßig gesetzten Stahlstützen ruht, sind die Wandscheiben frei in den Raum gestellt. Die großen Glaswände lassen Innen- und Außenraum verschmelzen; luxuriöse Materialien wie eine Wand aus golden schimmerndem Onyxmarmor und das Farb- und Lichtspiel aus verchromtem Stahl, schwarzem Teppich, weißen Ledersesseln, roten Vorhängen und dem Wasser der Außenbassins schaffen eine noble Eleganz. 1986 wurde der nach der Ausstellung abgerissene Pavillon, der seitdem als Mythos durch die Architekturgeschichte geisterte, rekonstruiert.

Zur gleichen Zeit baute Mies für das Ehepaar Tugendhat im tschechischen Brünn eine weiß verputzte Luxusvilla in Hanglage. Das

tually be called the most beautiful building of the 20th century." For the first time, Mies implemented his ideal of free-flowing space. The wall slabs are freely arranged under the flat roof which appears to hover in the air and is supported by eight regular steel columns. The large glass walls cause the interior and exterior to merge; luxurious materials such as a wall of gold-glistening onyx marble and the interplay of colour and light in the chrome-plated steel, black carpet, white leather armchairs, red curtains and the water of the outdoor pond create a noble elegance. After the exhibition the pavilion was pulled down, but it survived as a legend in architectural history and was reconstructed in 1986.

At the same time, Mies built a luxury hillside villa with white external rendering in the Czech town of Brünn (Brno) for the

Haus Tugendhat in Brünn/Brno. Vorn der „Barcelona-Sessel"

Haus Tugendhat in Brünn/Brno. At the front the "Barcelona armchair"

untere Geschoss, der Hauptwohnbereich, besteht aus einem einzigen großen Raum, der nur durch eine halbrunde Ebenholzscheibe, eine Onyxwand und die in regelmäßigen Abständen stehenden verchromten Stahlstützen gegliedert wird. Die ins Tal gerichtete Außenwand ist in eine große Glasfläche aufgelöst.
Für alle drei Projekt entwarf Mies auch das Mobiliar, angeregt vermutlich durch seine neue Lebensgefährtin und enge Mitarbeiterin Lilly Reich, eine Innenarchitektin und Modeschöpferin, die später Mies' Berliner Büro bis zu ihrem Tod 1947 weiterführte. In Stuttgart stellte Mies den frei schwingenden MR-Stuhl vor, der allein aus einem gebogenen Stahlrohr, einem Ledersitz und einer Rückenstütze besteht. 1929 präsentierte er den „Barcelona-Sessel", eine Stahlkonstruktion mit gesteppten Kissen, die zu einer Ikone des modernen Designs wurde.

Tugendhat family. The lower floor, which is the main living area, consists of a single large room which is only sub-divided by a semi-circular ebony panel, a wall of onyx and the regularly placed chrome-plated steel columns. The outer wall overlooking the valley is a large expanse of glass.
For all three projects, Mies also designed the furniture, probably at the suggestion of his new female friend and close associate, Lilly Reich, an interior architect and fashion designer who later managed Mies' office in Berlin until her death in 1947.
In Stuttgart Mies presented the cantilever chair which consisted only of a curved tubular steel frame, a leather seat and a backrest. In 1929 he presented the "Barcelona chair", a steel structure with fabric cushions which became an icon of modern design.

DIE DREISSIGER JAHRE

1930 übernahm Mies die Leitung der berühmten, damals von politischen Grabenkämpfen geschwächten Kunstschule Bauhaus in Dessau, die er praktisch zur Architektenschule umwandelte. Hier wie später am Illinois Institute of Technology richtete Mies den Lehrplan auf das Studium der Bautechnik, der Materialien und der grundlegenden Bautypen aus und versuchte so, seinen Studenten „ein Gefühl für Proportion und Raum zu geben".

Im Oktober 1932 musste das Bauhaus aus politischen Gründen schließen, durfte aber als Privatinstitut weitergeführt werden. Nach einem halbjährigen Intermezzo in einem alten Fabrikgebäude in Berlin-Steglitz (Siemensdamm/Ecke Birkbuschstraße, kriegszerstört) beschloss die Lehrerschaft 1933 nach erneutem politischem Druck die Selbstauflösung.

Dennoch versuchte Mies zunächst, sich mit den Nationalsozialisten zu arrangieren. Er wurde eingeladen zum Reichsbank-Wettbewerb in Berlin und entwarf einen Vorschlag zum Deutschen Pavillon auf der Weltausstellung in Brüssel 1935. Doch Aufträge erhielt er keine mehr. Oder sie ließen sich nicht mehr verwirklichen – wie das Haus Ulrich Lange in Krefeld, das die Baubehörden als „undeutsch" ablehnten. Der Weg, den er mit dem Barcelona-Pavillon eingeschlagen hatte, wurde ihm versperrt.

Immer noch zögerte der bedächtige, des Englischen nicht mächtige Mies, Deutschland zu verlassen. Erst 1937 nahm er einen Bauauftrag in den USA an, den ihm sein eifrigster Fürsprecher, der junge Architekt Philip Johnson, verschafft hatte. 1932 hatte Johnson am New Yorker Museum of Modern Art die Ausstellung „International Style" mitorganisiert, die Mies auch in Amerika weithin bekannt gemacht hatte. In Chicago bot man ihm die Leitung der Architekturabteilung am Armour Institute an, und Mies schlug ein. 1938 verließ er Deutschland für immer.

DIE AMERIKANISCHEN JAHRE

Zwischen 1938 und 1969 konnte Mies wesentlich mehr Projekte verwirklichen als in den dreißig Jahren zuvor. Neben seiner Professur unterhielt Mies stets ein eigenes Büro in der Innenstadt von Chicago. Gewohnt hat er interessanterweise in einem Neo-Renaissance-Haus, allein, obwohl er fast 30 Jahre lang mit der Bildhauerin Lora Marx liiert war. 1944 nahm er die amerikanische Staatsbürgerschaft an.

Für seine bald zum Illinios Institute of Technology (IIT) fusionierte Hochschule projektierte Mies einen neuen Campus, der in Etappen entstehen sollte. Über das lang gestreckte Gelände spannte Mies ein Raster, an dem er die rechteckigen Labor- und Wohnge-

THE 1930s

In 1930 Mies became the director of the world-famous Bauhaus School of Art in Dessau, which was at the time weakened by political in-fighting, and practically converted it to a School of Architecture. Here, as in his later work at the Illinois Institute of Technology, Mies based the curriculum on the study of building technology, materials and the fundamental types of buildings, aiming to give his students "a feeling for proportions and space".

In October 1932 the Bauhaus had to close for political reasons, but it was allowed to continue as a private institute. After a six month intermezzo in an old factory building in Berlin-Steglitz (on the corner of Siemensdamm and Birkbuschstrasse, destroyed in the war) the teaching staff decided to dissolve the institution in 1933 under renewed political pressure.

Nevertheless, Mies initially tried to come to an accommodation with the National Socialists. He was invited to participate in the Reichsbank competition in Berlin and he designed a proposal for the German pavilion at the International Exhibition in Brussels in 1935. But he did not receive any more commissions. Or they could no longer be implemented – like the Ulrich Lange house in Krefeld which the building authorities rejected as being "un-German". He was prevented from continuing the path that he had begun in the Barcelona pavilion.

Mies still hesitated to leave Germany because he was reticent by nature and unable to speak English. It was not until 1937 that he accepted a construction commission in the USA which his keenest advocate, the young architect Philip Johnson, had obtained for him. In 1932 Johnson was one of the organisers of the exhibition "International Style" at the New York Museum of Modern Art, which had also made Mies known in America. In Chicago Mies was invited to direct the Department of Architect at the Armour Institute, and he accepted. In 1938 he left Germany for ever.

THE AMERICAN YEARS

Between 1938 and 1969 Mies was able to implement many more projects than in the previous thirty years. In addition to his professorial duties, Mies always had his own bureau in the centre of Chicago. Interestingly, he lived in a neo-Renaissance house – alone, although he had a liaison with the sculptress Lora Marx which lasted almost 30 years. He became an American citizen in 1944.

For his college, which was soon merged to form the Illinois Institute of Technology (IIT), Mies designed a new campus which was to be built in stages. Mies created a grid over the elongated land area, and aligned the rectangular laboratory and residential build-

bäude ausrichtete. Wie in seinen Wohnhäusern zielte er auch hier im Großen darauf ab, fließende Räume zu schaffen, die weder einengend noch uferlos wirken. Auch den einzelnen Bauten liegt ein Modul zugrunde. Es sind zumeist zwei- oder dreigeschossige Flachbauten mit sichtbarem Stahlskelett, das mit Glas oder Backstein ausgefacht ist. Die einzelnen Bauten ähneln einander, doch verfeinerte Mies beständig sein Gespür für Stahl; auf die Wahl der Profile und Eckbildungen verwandte er größte Sorgfalt.

1946-51 entstand westlich von Chicago das reduzierteste und wohl umstrittenste Wohnhaus des 20. Jahrhunderts: das Haus Farnsworth. Mitten in ein großes Waldgrundstück, nicht einsehbar von den Nachbarn, setzte Mies einen schlichten Kasten aus Glas und weiß lackiertem Stahl. Acht Stützen tragen die Bodenplatte, die 1,50 Meter über dem Erdreich schwebt, und die Decke. Die Wände sind vollkommen verglast. Ein „Servicekern" im Innern verbirgt zwei Badezimmer, eine Schrankwand trennt Wohn- und Schlafbereich.

Ende der 40er Jahre erbaute Mies am Ufer des Michigansees in Chicago die ersten Wohnhochhäuser überhaupt, deren Außenhaut nur aus Stahl und Glas besteht. 1954-58 folgte das Seagram-Building in New York, das rasch zum Prototyp der modernen Büroarchitektur in aller Welt wurde.

Seine Vorstellung von einem „universalen" Raum verwirklichte Mies erstmals in der Crown-Hall auf dem IIT-Campus, einem rechteckigen Glas- und Stahlbau von 36,5 Metern Spannweite, dessen Inneres völlig frei aufteilbar ist. Unverwirklicht blieb der Entwurf einer Convention Hall für Chicago, einer auf dem Quadrat basierenden Mehrzweckhalle, die mit einer Seitenlänge von 220 Metern die größte der Welt geworden wäre. Sein Lebenswerk krönte Mies schließlich mit seinem wohl abstraktesten Bau, der Neuen Nationalgalerie in Berlin.

Ein Jahr nach deren Eröffnung, am 17. August 1969, starb Ludwig Mies van der Rohe in Chicago, beigesetzt wurde er auf dem dortigen Graceland Cemetery.

MIES' ARCHITEKTURPHILOSOPHIE

Geprägt durch eine handwerklich-technische Erziehung, setzte sich Mies erst spät, dafür umso intensiver mit Philosophie auseinander. Mies glaubte an die „objektive" Wahrheit, die sich in der Architektur ausdrücken müsse. Es galt die Form zu „finden", nicht zu „erfinden". Folgte er anfangs Hendrik Berlage, der den Künstler als „Dolmetscher" des Zeitgeists definiert hatte, so strebte er später einen Ausgleich zwischen dem „mechanistischen Ordnungsprinzip, das die materiellen und funktionellen Tendenzen überbe-

ings on this grid. As in his residential houses, Mies aimed to create flowing space on this larger scale which would not appear restrictive or unbounded. The individual buildings are also based on a modular principle. They are mainly relatively low two or three storey buildings with a visible steel framework and glass or brick in the spaces between the framework elements. The individual buildings are similar to each other, but Mies constantly developed his feeling for steel; he took great care in the selection of profile sections and the design of the corners.

From 1946-51, a house arose to the west of Chicago which was probably the most reduced and controversial dwelling house of the 20th century: the Farnsworth house. In the middle of a large woodland plot which is not overlooked by the neighbours, Mies set a plain box structure of glass and white painted steel. Eight columns bear the floor slab, which is 1.50 metres above the ground, and the ceiling. The walls are completely glazed. A "service core" in the interior contains two bathrooms, and a fitted cupboard separates the living and sleeping areas.

At the end of the 1940s Mies built residential tower buildings with an outer facade entirely of steel and glass by the shore of Lake Michigan in Chicago - the first ever residential buildings of this type. In 1954-58 they were followed by the Seagram building in New York, which quickly became the prototype of modern office architecture throughout the world.

Mies first implemented his concept of a "universal" room in the Crown-Hall on the IIT campus, an octagonal glass and steel building with a span of 36.5 and an interior that can be sub-divided with complete flexibility. His design for a Convention Hall for Chicago was never built. It was to be a multi-purpose hall on a square ground plan with sides of 200 metres, which would have made it the largest in the world. Mies finally crowned his life's work with what was probably his most abstract building, the New National Gallery in Berlin.

A year after it opened, on 17th August 1969, Ludwig Mies van der Rohe died in Chicago and was buried there in Graceland Cemetery.

MIES' ARCHITECTURAL PHILOSOPHY

In view of his craft-oriented and technical education Mies only turned to Philosophy later on, but did so all the more intensively. Mies believed in the "objective" truth which should express itself in architecture. The task was to find the form, not to invent it. Initially he followed Hendrik Berlage, who had defined the artist as an "interpreter" of the spirit of the age, but later he strove for reconciliation between the "mechanistic principle which over-

tont", und dem „idealistischen Ordnungsprinzip" mit seiner „Überbetonung des Ideellen und Formalen" an. Kreativität erwies sich für Mies nicht im intuitiven Schöpfungsakt, sondern im geduldigen Aufspüren der „Wahrheit" und deren Umsetzung in Raum und Form. Gern und oft zitierte er Augustinus' Wort: „Schönheit ist der Glanz der Wahrheit."
Mit Beharrlichkeit und Geduld verfolgte Mies seine Ziele und Visionen. Tage und Wochen konnte er an Modellen verbringen und an Details feilen – bis er das Bauproblem auf die Quintessenz verdichtet hatte. Entwerfen verstand er als die Kunst der Reduktion auf die materialgerechte und technisch optimale Form. „Baukunst fängt mit dem sorgfältigen Zusammenfügen zweier Ziegelsteine an." Und Mies besaß ein untrügliches Gespür für Materialien, Proportionen und Raum.
Interessanterweise schaffen Mies' rigide Schemata eine „Ordnung, die dem Leben freien Spielraum zur Entfaltung lässt". Mies' Ziel war der „behütende", nicht der „umschließende" Raum – zunächst in dynamischem Fluss wie im Barcelona-Pavillon, später als statisch-symmetrischer Einheitsraum, als „neutraler Rahmen, in dem Menschen und Kunstwerke ihr eigenes Leben führen können".
Folgerichtig präsentierte Mies für unterschiedliche Aufgabenstellungen oft gleiche Lösungen. In seinen späteren Jahren beschränkte er sich auf nur noch zwei Bautypen, die er variierte und verfeinerte: das Hochhaus und den Pavillon als universalen Raum. Dabei steht seine Architektur nie isoliert, sondern ist immer im Zusammenspiel mit der Umgebung zu verstehen. „Wir sollten uns bemühen, Natur, Häuser und Menschen zu einer höheren Einheit zusammenzubringen."
Mies war ein Mann fester Überzeugungen, die er schnörkellos auf den Punkt brachte – wie mit seinem berühmten Aphorismus „Weniger ist mehr". Sich „hinter geistreichen Theorien zu verstecken", lehnte er ab.
„Was Mies einen solchen Einfluss verschafft, ist der Umstand, dass er so leicht zu kopieren ist", sagte Philip Johnson in den 60er Jahren. Doch eines hat sich gezeigt: Fast allen Kopien fehlt, was die Miesschen Originale so faszinierend macht: das bezwingende Gefühl, so und nicht anders sei es „richtig".
Mies' Bauten umgibt eine Aura des Ewiggültigen, der man sich nur schwer entziehen kann. In ihrer Klarheit und Einfachheit, ihrer Konsequenz und Präzision wirken sie wie die Körper gewordenen ewigen Grundprinzipien der Architektur, zeitlos erheben sie sich über Stile und Moden – und über die Persönlichkeit ihres Architekten.

emphasises the material and functional tendencies", and the "idealistic principle" with its "over-emphasis on ideals and formal questions". Mies believed that creativity lay not in an intuitive creative act, but in the patient discovery of the "truth" and its realisation in space and form. He often and gladly quoted Augustine: "beauty is the splendour of truth."
Mies pursued his goals and visions with persistence and patience. He could spend days and weeks working on models and perfecting details – until he had reduced the construction problem to its quintessence. He regarded design as the art of reduction to a form that does justice to the materials and is technically optimised. "Architecture begins with the careful joining of two bricks." And Mies had an infallible feeling for materials, proportions and space.
Interestingly, Mies' rigid principles created an "order which gives life free scope for development." Mies' goal was the "sheltering" room, not the "enclosing" room – initially in a dynamic flow as in the Barcelona pavilion, later as a static and symmetrical universal room, a "neutral framework within which people and works of art can lead their own life."
Consequently, Mies often presented the same solutions for different tasks. In his later years he limited himself to just two types of building which he varied and developed: the tower building and the pavilion as a universal room. But his architecture is never to be seen in isolation, it must always be understood in its interaction with its surroundings. "We should strive to bring nature, buildings and people together in a higher unity. When you look at nature through the glass walls of the Farnsworth house, it has a deeper meaning than when you are standing outside."
Mies was a man of firm opinions which he formulated succinctly and without any frills, as in his famous aphorism "less is more". He refused to "hide behind ingenious theories".
"What gives Mies such an influence is the fact that he is so easy to copy", Philip Johnson said in the 1960s. But almost all copies lack what makes the originals by Mies so fascinating : the overpowering feeling that they are "just right" as they are and could not be any different.
Mies' buildings have an aura of eternal validity which it is difficult to escape from. In their clarity and simplicity, their consistency and precision they seem to be embodied prototypes of the eternal principles of architecture, they rise timelessly above styles and fashions – and above the personality of their architect.

BAUTEN UND PROJEKTE FÜR BERLIN UND POTSDAM

REALISIERTE BAUTEN

Haus Riehl

Potsdam-Neubabelsberg, Spitzweggasse 3, 1907
Bruno Paul sagte über das Werk seines 20-jährigen Mitarbeiters Ludwig Mies: „Das Haus hat nur einen einzigen Fehler: dass ich es nicht gebaut habe."
Das Landhaus für den Philosophieprofessor Riehl ist ein Bau mit zwei Gesichtern. Mies setzte es nicht in die natürliche Hanglage, sondern in einen ebenen Rosengarten, der durch eine Mauer nach unten abgestützt wird. Oben, vom Hauptzugang aus, stellt es sich als schlichtes Landhaus konservativen Zuschnitts dar: als verputzter Backsteinbau mit Dachgeschoss, dessen Eingang in

BUILDINGS AND PROJECTS FOR BERLIN AND POTSDAM

IMPLEMENTED BUILDINGS

Haus Riehl

Potsdam-Neubabelsberg, Spitzweggasse 3, 1907
Bruno Paul commented on the work by his 20 year old staff member Ludwig Mies: "The house has only one fault: that it was not built by me."
The country house for the Philosophy professor Riehl is a building with two faces. Mies did not set it into the natural slope, but rather into a level rose garden which is supported below by a wall. From the main entrance at the top it looks like a plain country house of a conservative type: a rendered brick building with an attic, with its entrance situated under a gable in the middle of

Haus Riehl, Eingangsfront

Haus Riehl, entrance facade

der Mitte der Längsseite unter einem Giebel sitzt. Anders die Ansicht vom Fuß der Anhöhe: Über der breiten Stützmauer erhebt sich die Stirnseite wie eine klassische Giebelfront. Das Untergeschoss öffnet sich zum Hang, das Erdgeschoss ist in eine offene Veranda aufgelöst, deren Stützen den Giebel tragen. Sie ist heute durch Fenster geschlossen.
Die Qualitäten des Hauses bemerkt man erst bei genauerem Hinsehen: Feine Pilaster, Eckpfosten und ein Gebälk bilden ein konstruktives Gerüst, in das die Wandflächen straff gespannt sind.
Der Grundriss ist symmetrisch aufgebaut: Durch die Eingangsdiele betritt man eine zentrale Halle, an die sich rechts die Veranda und zwei kleine Zimmer, links das Wohnzimmer und ein Arbeitszimmer anschließen. Die Küche und die Arbeitsräume befinden sich im Souterrain, das zum Hang hin belichtet wird. Im Dachgeschoss sind vier Schlafzimmer untergebracht. Die originale Innenausstattung war nüchtern und sachlich. Mies' Erstling, von den Riehls „Klösterli" genannt, überrascht durch seine Strenge, seine erstaunliche Reife und Ausgeglichenheit.

Haus Perls mit Anbau (links)
Haus Perls with extension (left)

the long side. The view from the foot of the hill (now Karl-Marx-Strasse) is different: the end of the house rises like a classical gable facade above the broad supporting wall. The lower storey opens out to the slope, the ground floor forms an open veranda with columns which support the gable. The veranda has now been closed by windows with small sections.
The finer qualities of the house are only seen on closer inspection: fine pilasters, corner posts and timberwork form a structural framework into which the walls are tightly fitted.
The floor plan is symmetrical: the entrance hall leads into a central room which has the veranda and two small rooms to the right, and the living room and a study to the left. The kitchen and the working rooms are in the basement which is illuminated from the slope. The attic contains four bedrooms. The original interior was plain and rational, the walls were clad up to the door lintels. Mies' first project, which the Riehls called their "little monastery", surprises us with its austerity, maturity and poise.

Haus Perls

Berlin-Zehlendorf, Hermannstraße 14, 1910-11, als Haus Fuchs 1928 um einen Galerietrakt erweitert
1910, während seiner einjährigen Auszeit vom Büro Behrens, entwarf Mies für den wohlhabenden Rechtsanwalt und Schinkel-Liebhaber Hugo Perls ein streng klassizistisches Landhaus. Es ist ein zweigeschossiger Putzbau, über dessen auffälligem Gesims sich ein flach geneigtes Walmdach erhebt. Der bis auf die Fensterläden völlig schmucklose Bau steht quer zur Straße. Zum Garten hin öffnet sich die streng symmetrische, fünfachsige Hauptfassade in einer tiefen, durch quadratische Pfeiler dreigeteilten Loggia – eine Reminiszenz an Schinkels Neuen Pavillon am Schloss Charlottenburg.
Hinter der Loggia lag im Zentrum des Hauses das große Speisezimmer, das Max Pechstein wenig später mit Wandbildern ausmalte. Von hier gingen zur einen Seite die Bibliothek und das Musikzimmer, zur anderen das Arbeitszimmer ab.

Haus Perls

Berlin-Zehlendorf, Hermannstrasse 14, 1910-11, gallery added in 1928 when it was known as Haus Fuchs
In 1910, during his one year leave from the Behrens office, Mies designed a strictly Classical country house for the wealthy solicitor and Schinkel enthusiast Hugo Perls. It is a two storey rendered building with a shallow sloping hipped roof above the striking cornice. The building is without ornamentation except for the window shutters and stands at right angles to the road. On the side facing the garden, the strictly symmetrical main facade of five axes opens out to a deep loggia divided into three parts with square pillars – reminiscent of Schinkel's new pavilion at Charlottenburg Palace.
Behind the loggia in the centre of the house was the large dining room to which Max Pechstein added wall paintings soon afterwards. The library and music room led off on one side, the study on the other.

Das Haus Perls strahlt eine klassische Würde aus und wirkt doch nicht so schwergewichtig wie das Haus Wiegand, das Peter Behrens 1911-12 in Dahlem errichtete (Peter-Lenné-Str. 28-30).
Für den späteren Besitzer Eduard Fuchs, einen bekannten Kulturhistoriker, baute Mies 1928 einen Ausstellungstrakt an. Die Proportionen und die Fensterformen sind dem alten Bauteil angepasst, das Flachdach ist dagegen ein Element der 20er Jahre.

Haus Werner

Berlin-Zehlendorf, Quermatenweg 2-4, 1912-13
Mies' erstes Haus nach dem Sprung in die Selbstständigkeit wurde sein konventionellstes. Mit dem Mansarddach, den Gauben und den kleinteiligen Fenstern mit Außenblenden erinnert der

Haus Perls has a classical dignity, but it is not as monumental in ist appearance as Haus Wiegand, which Peter Behrens built in Dahlem 1911-12 (Peter-Lenné-Str. 28-30).
For the later owner Eduard Fuchs, a well-known culture historian, Mies added an exhibition wing in 1928. The proportions and the window forms are based on the old part, but the flat roof is a typical element of the 1920s.

Haus Werner

Berlin-Zehlendorf, Quermatenweg 2-4, 1912-13
Mies' first house as a freelance architect was also his most conventional building. With its mansard roof, dormers and small windows with external blinds, the rendered brick building is reminis-

Haus Werner, Gartenansicht

Haus Werner, view from the garden

verputzte Backsteinbau an barocke Landhäuser. Zur Straße hin ist die Symmetrie durch einen vorgezogenen Bauteil (mit der Küche) aufgebrochen. Seit dem An- und späteren Ausbau der Gärtnerwohnung in den 20er Jahren stellt sich das Haus Werner als zweiteilige Baugruppe dar.
Eindrucksvoller ist die Rückansicht, die sich zu einer fast barocken Gartenanlage mit Säulengang öffnet. Die symmetrische Gartenfront wird dominiert durch den zweigeschossigen, übergiebelten und leicht vorgezogenen Mittelteil mit Fenstertüren.
Der Grundriss folgt einem sechsteiligen Schema, das Mies ab dem Haus Werner gern verwandte: Seitlich der Eingangshalle liegen zur Straße hin rechts der Salon, links die Treppe und die Küche. Zum Garten hin liegt hinter der Halle das Wohnzimmer, rechts das Arbeits-, links das Speisezimmer. Die Schlafzimmer sind im voll ausgebauten Dachgeschoss untergebracht. Für die Inneneinrichtung entwarf Mies Möbel im neoklassizistischen Stil.

Haus Urbig

Potsdam-Neubabelsberg, Virchowstraße 23, 1915-17
Die großzügige neoklassizistische Villa, die Mies für den Bankier Fritz Urbig oberhalb des Griebnitzsees errichtete, wurde in den 90er Jahren mustergültig renoviert. Dabei handelt es sich um einen zweigeschossigen rosaroten Putzbau mit ausgebautem Dachgeschoss. Zur linken Schmalseite springen aus dem rechteckigen Baublock die Küche, nach hinten der Speisesaal vor, dessen Dach als große Terrasse angelegt ist. Interessant ist der Grundriss der Gesamtanlage, denn das Haus ergänzt sich mit der vom Speisesaal eingefassten Gartenterrasse zum Quadrat.
Die symmetrische Vorderfront erhält ihren Rhythmus durch flache Wandvorlagen. Die Fenster sind mit Travertin eingefasst und wirken durch schmale Fugen wie aus der Wandfläche ausgeschnitten. Über den Fenstertüren im Erdgeschoss sitzen aufwändige Schuckreliefs. Die Mitte der Fassade ist durch den rundbogigen Eingang und einen Balkon im ersten Stock ausgezeichnet. Auf den drei übrigen Seiten fehlen die Pilaster; Fensterläden verleihen dem Bau hier eine gewisse Leichtigkeit.
Während das Haus nach vorne hin auf einem dezenten Sockel steht, präsentiert sich die Rückfront majestätisch über mehreren mit Treppen verbundenen Terrassen. Der Garten, der ursprünglich sanft bis hin zum Seeufer abfiel, wurde durch den Mauerbau gekappt.
Das Innere der Villa ist klassizistisch kühl, fast pompös gehalten. Von der Eingangshalle erreicht man durch zwei dorische Säulen hindurch - ein Motiv aus Schinkels Schloss Tegel - einen querlie-

cent of Baroque country houses. On the side facing the street, the symmetry is broken up by an extended front section (with the kitchen). Since the gardener's flat was added and later extended in the 1920s, Haus Werner has been an ensemble made up of two elements.
The view from the rear, which opens up to an almost Baroque garden with a colonnade, is more impressive. The strictly symmetrical garden facade is dominated by the two storey, slightly projecting gabled central section with the French windows.
The floor plan has a six-part structure which Mies often used in designs after Haus Werner: next to the entrance on the street side, the drawing room is on the right and the staircase and kitchen on the left. The living room is behind the hall looking out onto the garden, with the study on the right and the dining room on the left. The bedrooms are in the fully appointed attic. For the interior, Mies designed furniture in the neo-Classical style.

Haus Urbig

Potsdam-Neubabelsberg, Virchowstrasse 23, 1915-17
The generously sized neo-Classical villa which Mies built for the banker Fritz Urbig and his wife above Griebnitzsee during the First World War was admirably renovated in the 1990s. It is a two-storey building with pink rendering, with a fully fitted attic six axes wide and three axes deep. By the left narrow side, the kitchen extends out from the rectangular main building and the dining room extends to the rear, with a large terrace on its roof. The ground layout of the entire complex is interesting, because the house and the garden terrace enclosed by the dining room form a square.
The symmetrical front facade has a rhythmic pattern due to the low front walls. The windows are lined with travertine, and the narrow joints make them appear to be cut out of the wall surface. There are intricate decorative reliefs over the glazed doors on the ground floor. The centre of the facade is characterised by the round arch of the entrance and a first floor balcony. On the other three sides there are no pilasters; window shutters give the building a certain lightness.
The front of the building rests on an unobtrusive pedestal, but the rear has a majestic appearance with several terraces linked by flights of steps. The garden, which originally fell away gently to the bank of the lake, was cut off when the Wall around West Berlin was built.
The interior of the villa is classically detached, almost pompous. The entrance hall leads between two Doric columns - a motif

Haus Urbig, Gartenseite

Haus Urbig, side facing the garden

Haus Mosler, Straßenseite

Haus Mosler, side facing the street

genden Korridor. Geradeaus lag das Musikzimmer, rechts das Herren- (nach vorn) und das Damenzimmer, links das Treppenhaus und der große Speisesaal.
Während der Potsdamer Konferenz diente das Haus Urbig Winston Churchill als Quartier, zu DDR-Zeiten wurde es als Gästehaus der Universität Potsdam genutzt. Von der Originalausstattung blieb wenig erhalten.

Haus Eichstädt

Berlin-Nikolassee, Dreilindenstraße 30, Ecke Lohengrinstraße, 1921-22
Dem ganz mit wildem Wein bewachsenen Haus sieht man Mies' Autorschaft nicht an. Für den Buchhändler Georg Eichstädt entwarf Mies ein zweigeschossiges verputztes Landhaus mit flach geneigtem Walmdach auf quadratischem Grundriss. Die Nordhälfte, auf der sich auch der Eingang befindet, beherbergt die Küche und die Diele, die Südhälfte wird von einem einzigen großen Raum eingenommen, der sich nach Osten halbkreisförmig erweitert. Hier stand ursprünglich ein Flügel.
Durch drei Fenstertüren öffnet sich dieser Wohn-, Speise- und Musikraum auf eine breite Terrasse, die sich nach Westen, an der Ecke des Hauses in einer überdachten Veranda fortsetzte. Das Haus wurde später nach Westen erweitert (am Dach zu erkennen), wobei die Veranda ins Gebäude mit einbezogen wurde. Der Garten war durch Pergolen und Hecken ins Gesamtensemble mit einbezogen.

from Schinkel's Tegel Palace - to a transverse corridor. Straight ahead was the music room, to the right the men's lounge (facing the front) and the ladies' lounge, on the left were the staircase and the large dining room.
During the Potsdam Conference Haus Urbig was used as the quarters for Winston Churchill; in the GDR period it was a guest house for the University of Potsdam. Little is preserved of the original fittings.

Haus Eichstädt

Berlin-Nikolassee, Dreilindenstrasse 30 (corner of Lohengrinstrasse), 1921-22
This house, which is covered in climbing wild vine, does not betray the authorship of Mies. For the bookseller Georg Eichstädt, Mies designed a two storey rendered country house with a shallow sloping hipped roof on a square floor plan. The northern half, which also includes the entrance, contains the kitchen and hall, the southern half contains a single large room which extends in a semi-circle to the east. There was originally a grand piano here.
This living, dining and music room opens out through three French windows onto a broad terrace which continued in a roofed veranda to the west at the corner of the house. The house was later extended to the west (as the roof shows), and the veranda was integrated into the building. The garden was integrated into the overall ensemble by pergolas and hedges.

Haus Feldmann

Berlin-Grunewald, Erdener Straße 2, 1921-22

Das Haus Feldmann wurde nach dem Zweiten Weltkrieg völlig überformt. Es ist im Kern ein zweigeschossiger, fünfachsiger (ursprünglich unverputzter) Backsteinbau mit Walmdach und niedrigeren Anbauten auf beiden Schmalseiten.

Haus Mosler

Potsdam-Neubabelsberg, Karl-Marx-Straße 28, 1924-26

Bei seinem letzten Landhaus im klassizistischem Stil folgte Mies in Grund- und Aufriss dem altbewährten Schema. Es ist ein lang gestreckter unverputzter Backsteinbau mit symmetrischer Fassade, hohem Walmdach und seitlichem niedrigerem Anbau. Die Details sind sorgfältig gearbeitet: beispielsweise die Fensterumrahmungen und das Hauptgesims aus Naturstein, der Fries unter dem Dachansatz, der Dekor der Balkonbrüstung oder die profilierten Backsteinstützen der rückwärtigen Loggia.

Das Innere ist wiederum sechsteilig aufgebaut, das Haupttreppenhaus liegt gleich links vom Eingang. In seiner asketischen Schlichtheit wirkt das Haus Mosler nobel und zugleich imposant.

Denkmal für Karl Liebknecht und Rosa Luxemburg

Friedhof Berlin-Friedrichsfelde, 1926 (zerstört)

Beim Historiker und KPD-Funktionär Ernst Fuchs, der das Haus Perls gekauft hatte, sah Mies den Entwurf eines Denkmals für die 1919 ermordeten Kommunisten-Führer und war entsetzt („ein hübsches Denkmal für einen Bankier"). Stattdessen schlug er Fuchs ein Monument in Form einer Mauer vor, „da die meisten dieser Leute vor einer Mauer erschossen wurden". Mies erhielt den Auftrag und entwarf eine kubische Skulptur aus rauhen, jedoch sorgfältig gefügten Backsteinen - eine monumentale Wand aus vor- und rückspringenden ungleichmäßigen Blöcken, sechs Meter hoch, zwölf Meter breit und vier Meter tief. Die Schrift in großen Lettern „Ich bin, ich war, ich werde sein" wurde vor 1931 beseitigt. Rechts prangte ein stählerner Stern von zwei Metern Durchmesser mit Hammer und Sichel und eine Fahnenstange. Das Denkmal wurde 1933 von den Nazis zerstört.

Wohnanlage Afrikanische Straße

Berlin-Wedding, Afrikanische Straße 15-41, 1926-27

Längs der Westseite der Afrikanischen Straße entwarf Mies vier ockerfarbig verputzte Blöcke mit Wohnungen zu zwei oder drei kleinen Zimmern, Küche und Bad. Drei Blöcke sind identisch, der

Haus Feldmann

Berlin-Grunewald, Erdener Strasse 2, 1921-22

Haus Feldmann was completely altered after the Second World War. In its core it is a two-storey, five-axis (originally unrendered) brick building with a hipped roof and lower out-buildings on both narrow sides.

Haus Mosler

Potsdam-Neubabelsberg, Karl-Marx-Strasse 28, 1924-26

In his last country house in the Classical style, Mies followed the established pattern for the ground layout and elevation. It is an elongated, unrendered brick building with a symmetrical five axis facade, a high hipped roof and a lower extension at the side. As always, the details are carefully executed: for example the window surrounds and main ledge of natural stone, the frieze under the roof, the decorative balustrade of the balcony, the sectional profile of the brick supports for the loggia at the rear. The interior consists of six sections with the main staircase to the left of the entrance. In its ascetic simplicity, Haus Mosler appears both elegant and impressive.

Monument for Karl Liebknecht and Rosa Luxemburg

Berlin-Friedrichsfelde cemetery 1926 (destroyed)

In the house of the culture historian and Communist Party functionary Ernst Fuchs, who had bought Haus Perls, Mies saw the design for a monument for the Communist leaders who had been murdered in 1919 and was horrified ("a pretty monument for a banker"). Instead, he suggested to Fuchs a monument in the form of a wall, "because most of these people were shot in front of a wall". Mies was given the commission and designed a cubic sculpture of rough but carefully jointed bricks - a monumental wall with irregular protruding and receding blocks which was six metres high, twelve metres wide and four metres deep. The text in large lettering: "I am, I was, I will be " was removed before 1931. On the right was a steel star two metres in diameter with a hammer and sickle and a flagpole. The monument was destroyed by the Nazis in 1933.

Residential complex, Afrikanische Strasse

Berlin-Wedding, Afrikanische Strasse 15-41, 1926-27

Along the western side of Afrikanische Strasse Mies designed four ochre-coloured rendered blocks with apartments consisting of two or three small rooms, kitchen and bathroom. Three blocks

Denkmal für Karl Liebknecht und Rosa Luxemburg (Foto von 1926)
Monument for Karl Liebknecht and Rosa Luxemburg (photograph: 1926)

Wohnanlage Afrikanische Straße
Residential complex, Afrikanische Strasse

vierte ist wesentlich kleiner, zeigt aber gleiche Formen. Durch Stichstraßen sind sie voneinander getrennt. Jeder der drei gleichen Blöcke besteht aus einem dreigeschossigen Haupttrakt und zwei zweigeschossigen Seitenflügeln, die so angefügt sind, dass sie wie nach hinten weggeklappt wirken. Dieser Eindruck wird durch die scharnierartigen runden Balkone in den Zwickeln - den einzigen Rundungen am ganzen Bauensemble - noch betont. Nur die Türverdachungen ragen aus den völlig planen Vorderfronten vor, während die Blöcke im Hof - dank der Balkone, des vorkragenden Daches und der lockeren Begrünung - fast intim wirken. Die Wohnungsgestaltung lässt sich an den Fenstergrößen der Vorderseite ablesen: ein Fensterflügel für das Bad, zwei für die „Kammer", drei für die „Stube". Die Häuserzeilen sind exakt um die Tiefe des Blockes von der Straße abgerückt; ursprünglich wurde jeder Eingang von zwei Pappeln - wie von zwei Säulen - gesäumt. Trotz des geringen Baubudgets wirken die Blöcke nicht billig, sie sind auch im Detail sehr sorgfältig gearbeitet.

are identical, the fourth is much smaller but has the same forms. They are separated from each other by access roads. Each of the three identical blocks consists of a three-storey main wing and two two-storey side wings which appear to be folded away to the rear. This impression is reinforced by the round hinge-like balconies in the joint between the wings - the only round element in the whole building complex. The front facades are completely flush, with only the door canopies projecting outwards, whereas the blocks in the courtyard appear almost intimate with the balconies, the overhanging roof and the informally arranged vegetation. The layout of the flats can be seen in the window sizes of the facade: one window element for the bathroom, two for the bedroom, three for the living room. The blocks are exactly one block depth away from the road; originally each entrance was framed by two poplar trees - like two columns. In spite of the low budget for the construction, the blocks do not appear cheap, in fact their details are very carefully executed.

Mit der Neuen Nationalgalerie schuf Mies van der Rohe die Idealform eines universell nutzbaren Stahl- und Glasbaus

With the New National Gallery, Mies van der Rohe designed the ideal type of a steel and glass building for universal use

Haus Lemke

Hohenschönhausen, Oberseestraße 56-57, 1932/33

Mies' letztes Wohnhaus, das er in Deutschland verwirklichen konnte, ist ein ganz schlichter, kleiner Flachbau aus rotem Backstein. Es ist das einzige realisierte Beispiel der Hofhäuser, mit denen sich Mies in den 30er Jahren intensiv beschäftigte. Der L-förmig angelegte, eingeschossige Bau wirkt zur Straße hin fast abweisend, öffnet sich aber in großen, bis zum Boden reichenden Glasflächen auf die Gartenterrasse. Zur Straße hin liegen die Küche, die Anrichte und ein „Mädchenzimmer", zum Garten hin das Wohnzimmer und eine „Halle", an die sich im hintersten Teil des Hauses das Schlafzimmer anschließt. Die Innentüren reichen bis zur Decke. Mit einbezogen in den Bau ist auch die Garage. 2001 wurde die Originalfassung weitgehend wiederhergestellt.

Neue Nationalgalerie

Tiergarten, Potsdamer Straße 50, 1962-68

Ab 1957 arbeitete Mies an einem Bürohaus für die Getränkefirma Bacardi in Santiago de Cuba, das aber nach Castros Revolution nicht mehr realisiert werden konnte. Als die Stadt Berlin an Mies herantrat, um ihn mit einem großen Bauauftrag - für die städtische „Galerie des 20. Jahrhunderts" - zu ehren, übertrug Mies diesen Entwurf auf das neue Projekt. Nach der Fusion der Galerie mit den West-Berliner Beständen der Nationalgalerie 1965 wurde das Gebäude 1965-68 als Neue Nationalgalerie errichtet.

Auf einem granitverkleideten Sockel erhebt sich die rundum verglaste Ausstellungshalle. Das weit vorkragende Stahldach wird allein von den acht Stützen außerhalb der Halle getragen - die beiden Installationsschächte im Innern sind nicht tragend. Die Halle misst 50 Meter im Quadrat und ist etwas über acht Meter hoch. Mit einer Seitenlänge von 64,80 Metern ist das Dach die größte jemals gefertigte starre Stahlplatte der Welt. Leichte Krümmungen in der Mitte der Platte und an den Ecken sorgen dafür, dass sie optisch nicht durchhängt.

Mies übertrug hier die Form eines klassischen Tempels in die moderne Sprache des Stahl- und Glasbaus. Die Decke ist kassettiert, die Kanten der Dachplatte sind in Felder geteilt, die Kugelgelenke, die die kreuzförmigen Stützen mit dem Dach verbinden, vertreten die klassischen Kapitelle.

Während die Glashalle für Sonderausstellungen genutzt wird, befinden sich die eigentlichen Ausstellungsräume wie auch die Büros im Sockel. Durch eine Mauer einbezogen ist ein Skulpturengarten im Westen, und die dort angrenzenden Räume erhalten Tageslicht.

Haus Lemke

Hohenschönhausen, Oberseestrasse 56-57, 1932/33

The last residential building which Mies' was able to build in Germany is a simple, small, low building of red brick. It is the only built example of the courtyard houses which Mies worked on intensively in the 1930s. The single-storey, L-shaped building appears almost defensive when seen from the street, but it opens up to the garden terrace with large windows which reach down to the floor. The kitchen, scullery and a "maid's room" are on the side facing the street, and on the garden side are the living room and a "hall" which leads to the bedroom at the rear of the house. The house has floor-to-ceiling interior doors. The garage is integrated into the house. After it had been disfigured by alterations, the original condition was largely restored up to 2001.

New National Gallery

Tiergarten, Potsdamer Strasse 50, 1962-68

From 1957 Mies worked on an office building for the Bacardi drinks company in Santiago de Cuba, but the project could not be built after Castro's revolution. When the city of Berlin approached Mies to honour him with a large building commission - for the municipal "Gallery of the 20th century" - Mies transferred this design to the new project. After the gallery was joined with the West Berlin works of the National Gallery in 1965, the building was built as the New National Gallery in 1965-68.

The exhibition hall with its all-round glazing stands on a granite-clad pedestal. The steel roof with its wide overhang is supported solely by the eight pillars outside the hall - the two installation shafts in the interior are not load-bearing structures. The hall is 50 metres square and slightly more than eight metres high. With its side length of 64.80 metres, the roof is the largest rigid steel panel ever made in the world. A slight curvature in the centre of the panel and the corners ensure that it does not visually "sag".

Here, Mies applied the form of a Classical temple to the modern style of a steel and glass building. The ceiling is coffered, the edges of the roof panel are divided into sections and the ball joints which link the cross-profile pillars with the ceiling correspond to Classical capitals.

Whereas the glass-clad hall is only used for special exhibitions, the main exhibition rooms and the offices are situated in the pedestal. A wall encloses a sculpture garden to the west, providing daylight to the adjacent rooms in the pedestal.